設計有約

inSIDE deSign

香港方黃建築師事務所地產項目空間設計系列專集 V · 方峻 著

華中科技大學出版社
http://www.hustp.com

分享设计
以之享受设计

Share Design Enjoy Design

Is originated by Mr. Fong

Unive

方峻

建築空間與多元跨界的中國香港設計師。先後在美國美聯大學、義大利米蘭理工學院、香港理工大學、國立華僑大學接受哲學、建築設計、設計管理的學士/碩士/博士等教育，同時也是香港室內設計協會專業會員、國際室內建築師設計聯盟會員、中國建築學會室內設計分會會員。其作品不但榮獲首屆國際建築景觀室內設計大獎賽金獎，還獲得過多項國內外室內設計的獎項與榮譽；更被美國《室內設計》和各類知名專業雜誌數次刊載。相繼出版且備受業界好評的個人設計專集分別有《"悟"設計》、《設計有約1》、《設計有約2》……《設計有約5》。

, a multidisciplinary designer in Hong Kong, China. Mr. Fong has successively received bachelor, master and doctor education in philosophy, architectural design and design management at Inter American University, Polytechnic University of Milan, The Hong Kong Polytechnic University and Huaqiao respectively. Also, he is a professional member of HK Interior Design Association, and a member of International Federation of Interior Architects/Designers and Institute of Interior Design of the Architectural Society of China. Besides the gold award gained at the 1st International Building Landscape Interior Design Awards (IBLIDA), his masterpieces have won numerous awards and honors in domestic and overseas interior design competitions. Moreover, his works have been published many times in various well-known professional magazines such as Interior Design in U.S.. In succession, he published his personal design collections including Inspiration Design, Inside Design I, Inside Design II, Inside Design V, etc.

Contents
目錄

- P006 信和・中央廣場銷售会所 Centre Squre Sales Office
- P022 信和・中央廣場示範單位 1 Centre Square Show Flat 1
- P034 信和・中央廣場示範單位 2 Centre Square Show Flat 2
- P046 信和・中央廣場示範單位 3 Centre Square Show Flat 3
- P058 信和・中央廣場示範單位 4 Centre Square Show Flat 4
- P068 南益・名仕華庭銷售會所 MingShiHuaTing Sales Office
- P084 南益・紫湖國際高爾夫別墅 1 Purple Lake International Golf Viila 1
- P108 南益・紫湖國際高爾夫別墅 2 Purple Lake International Golf Villa2
- P122 南益・紫湖國際高爾夫別墅 3 Purple Lake International Golf Villa 3
- P142 中海・世紀公館銷售会所 The Century Sales Office
- P156 中海・世紀公館別墅 1 The Century Villa 1
- P176 中海・世紀公館別墅 2 The Century Villa 2

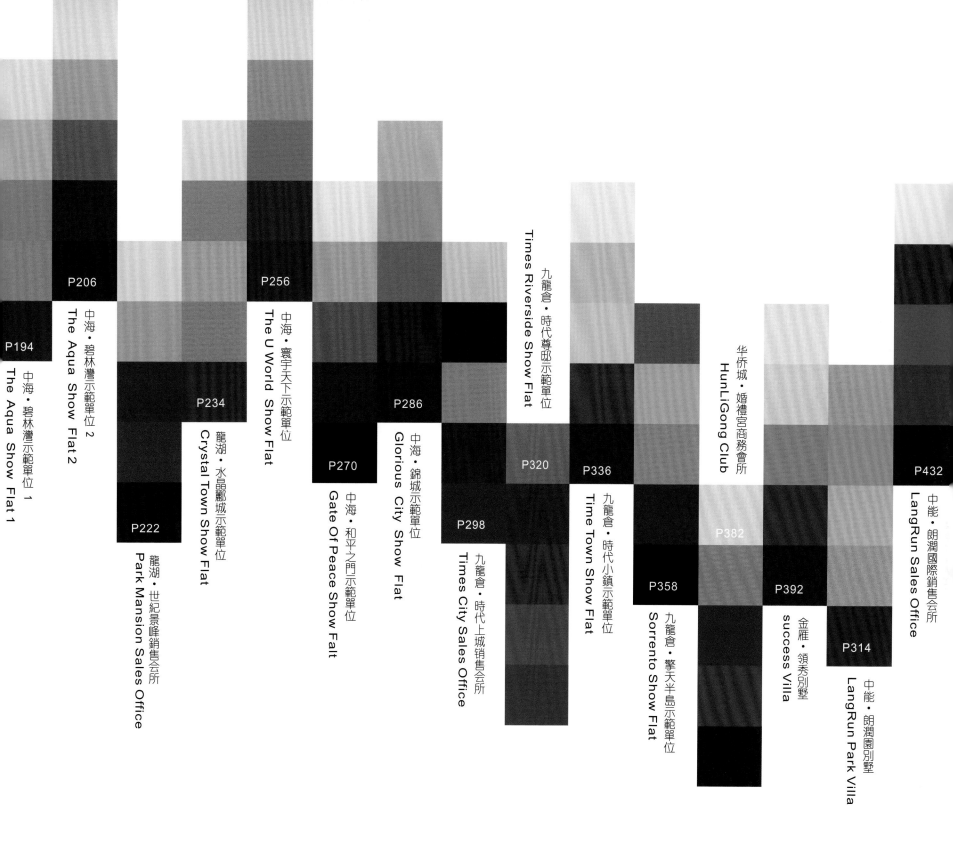

Centre Squre Sales Office

信和‧中央廣場銷售會所

Noble and conspicuous: white lotus.
In the eastern culture, the white lotus is acclaimed as the flower of purity for its graceful and elegant temperament. The brilliant whiteness of this flower is an optimum choice for enhancing the overall brightness of a picture, presenting a clean and clear vision. Apart from its quietness, which is described in an ancient Chinese poem as "washed by clean waves while showing no coquet", white also displays conspicuous nobility for its brilliance and pureness.

白莲花,高雅脱俗,被喻为圣洁之花。白莲花的白色明度很高,这种色彩是提高整体画面明度的极佳选择,给人以洁净清澈的视觉效果。除了"濯清涟而不妖"的雅静,白色因其高明度和高纯度,彰显出耀眼、夺目的华贵气质!

信和 • 中央廣場銷售會所

015 信和・中央廣場銷售會所

017 信和・中央廣場銷售會所

019 信和・中央廣場銷售會所

Centre Square Show Flat 1

信和・中央廣場示範單位 1

lively and elegant: pansy. The pansy is also named butterfly flower, and is rather popular in Europe for its colorfulness. Some white, some yellow, and some purple of the pansy create a lively picture. Refreshing white, brisk yellow, mysterious purple, all mix harmoniously with a stratified, pleasant and luxurious feeling, just appropriately interpreting the notion of longing.

三色堇,生动雅致。三色堇又名蝴蝶花,其色彩斑斓,在欧美颇受欢迎。几许纯白、几许微黄,几许紫色,三色堇的色调跳跃生动:白色纯净清爽、黄色愉悦轻松、紫色神秘优雅。这几种色彩的搭配,相得益彰,很有层次感,营造出愉快和华丽的氛围。作为表现"思慕"情怀的花卉,三色堇充满生机,情趣盎然。

Centre Square Show Flat 2

信和‧中央廣場示範單位 2

soft and light: Magnolia. Magnolia is a practically household name. Its full blossoms are so tender and fragrant on the trees. Pink purple petals set off elegant air of love. Pink purple is the absolute color of grace and softness. Pink may be regarded as childish, but when matched with potent colors, it could display a colorful and gorgeous adult world, and of course, a sweet harmonious and dreamlike atmosphere.

木兰花,柔和淡雅。几乎是家喻户晓的花卉。盛开在小乔木上的花儿粉妆玉琢,幽香四溢。粉紫色花瓣,散发着优雅的气息。在体现优雅、柔美时,粉紫色通常是当仁不让的色彩。粉色或许被认为是孩子气的颜色,但通过与一些强有力的色彩相搭配,也可以展现成人世界的艳丽华美,当然,亦可以营造出甜蜜祥和、宛如梦境般的氛围。

Centre Square Show Flat 3

信和・中央廣場示範單位 3

free and easy: herba orostachyos gray. Herba orostachyos generated in mountain slopes and rocks usually bears an elegant gray. Light gray, either velvet-like mouse gray or cold mountain-in-rain gray, thanks to its likeness to ground colors, could alone serve as a good foil to bright colors. In strong brightness, hazy gray could perfectly display a sense of softness and mystery. Gray also represents maturity and easiness to some extent, and tends to pose an impression of comfort and equilibrium.

瓦松灰,优雅洒脱。生于石质山坡和岩石上的瓦松,通常有一种很优雅的色泽,这是一种淡淡的灰白色,如天鹅绒般的鼠灰或是冷冷的烟山色,灰色,因为接近基本协调色,只要不包含其它色彩,都能很好地衬托出其他较为明亮的色彩。而在明度较高的情况下,灰色略带朦胧感的色调可以恰到好处地棒现出一丝轻柔和神秘。灰色在一定程度上还代表了成熟及洒脱,也更加倾向给予人舒适平衡的印象。

055 信和‧中央廣場示範單位3

Centre Square Show Flat 4

信和・中央廣場示範單位 4

quiet and exquisite: hydrangea. The hydrangea is a flower of various colors. Either white or purple, or blue delivers a feeling a cleanness and grace for its light colors. The umbrella-shaped blossom resemble snowballs, but without any sense of complication. White represents richness and exquisiteness. Its round and tightly clustered petals imply reunion.

绣球花,素净精致。绣球花是一种色彩多变的花。而因为色泽淡雅,无论白色、紫色还是蓝色的都能予人整洁优雅的感觉。白色绣球花,其伞形花序如雪球累累,却无繁复之感,白色在此意味着丰富与精致。其圆形和互相簇拥的花瓣,传递出美满团聚的寓意。

MingShiHuaTing Sales Office
南益·名仕华庭销售会所

Strelitzia Reginae, also known as Bird of Paradise flower, is named for its shape similar to red-crowned crane raising its head and looking out. Its scientific name is given in honor of Queen Charlotte, the wife of King George III of England. The special shape and color of Strelitzia Reginae present good quality. Apparently this color looks peaceful, with some indistinct dynamism, and the red vitality is partly hidden and partly visible. In addition, it also includes the carefree element of orange, creating a fashionable and leisure aura.

鹤望兰,又称天堂鸟或极乐鸟花。以形似仙鹤昂首远望而得名,其学名是为纪念英王乔治三世、王妃夏洛特皇后而取的。鹤望兰花形奇特、色彩给人感觉很有质地。表面看上去这种色彩很安定,隐约之中却又有几分动感,红色的活力若隐若现。另外又包含了橙色轻松愉快的元素,营造出时尚且悠然自得的气氛。

075　南益·名仕華庭銷售會所

南益·名仕華庭銷售會所

Purple Lake International Golf Villa 1

南益・紫湖國際高爾夫別墅 1

Sunshine pours onto tree branches and flowers, and it is so quiet as if you can hear the branches touching the shadow. Just like the silence before the opening of a good show, what you see after opening the door is another scene with both movement and tranquility in the courtyard. Flourishingly, the flowers in full bloom are so splendid! Bright-colored carpet, gorgeous flower arrangement in the vase, well-proportioned picture frames in different shapes on the wall, and similarly bright and flowery curtain rich multicolored decorations are in dizzying quantities for your eyes to see, alive with warm and unrestrained summer love.

阳光倾洒在树枝上与花丛中，似乎静谧得听得见枝头光影碰触的声响。就像一幕好戏开场前的静寂，推门所见，是另一番与庭院动静相宜的场景，繁华似锦。色彩斑斓的地毯，花瓶上绚烂的插花，墙壁上错落有致、形态各异的图框画框，色泽同样明丽香艳的窗帘------花团锦簇，令人目不暇接。洋溢着热烈奔放的夏日情怀。

Purple Lake International Golf Villa 2

南益·紫湖國際高爾夫別墅 2

luxurious and abundant: pot marigold. The name originates from the Calendula officinalis of Southern Europe, and is a traditional color in Britain and North America. As a potent and warm yellow, it reminds one of gold, for which the pot marigold symbolizes abundance, glory and beauty. The incorporation of the disposition of red strengthens its impact and sense of happiness.

金盏菊: 奢华富丽。这个色彩的名称源自原产于南欧的金盏花,在英美是很受欢迎也很传统的色彩,它是强而温暖的黄色,能让人联想到黄金。所以金盏花象征着丰富、光辉和美丽。因为融合了红色的个性,所以冲击力强,也更突显了黄色的幸福感。

abundant and beautiful: grain yellow. Nothing can better present a sensation of a bumper harvest than grain gold. The combination of abundance, glory and beauty brings an intense feeling of happiness. Also known as sun-orange, this color is associated with the harvest season because the orange is said to be a symbol of fertility since the Rome times. In practice, it embodies wealth and happiness, which makes it a suitable color for family style.

稻谷金,丰盈美好。没有什么色彩能比稻谷金更能呈现丰收之感了。丰富、光辉和美的结合,使这种色彩给人带来强烈的幸福感。这种色亦为"SUN-ORANGE",据说在欧洲,橙子的果实从罗马时代开始就被当作多产的象征,因此从这个色彩就很容易令人联想到收获季节。在运用中,它象征着富贵和幸福,也是适合家庭系的印象色彩。

The Century Sales Office

中海·世紀公館銷售会所

There is luxury in simplicity, and cheerfulness in quietness. The green is partly hidden and partly visible in the space, generating invisible attraction and affinity. The compact and orderly layout seems random and ornaments are delicate and unconventional. Each detail is worth thinking.

平实中带点华贵,沉静中又有些许轻快。绿意在空间中若隐若现,会所具有种无形的吸引力、亲和力。布局紧凑有序,看似随意的摆件精致不落俗套,每一个细节都耐人寻味。

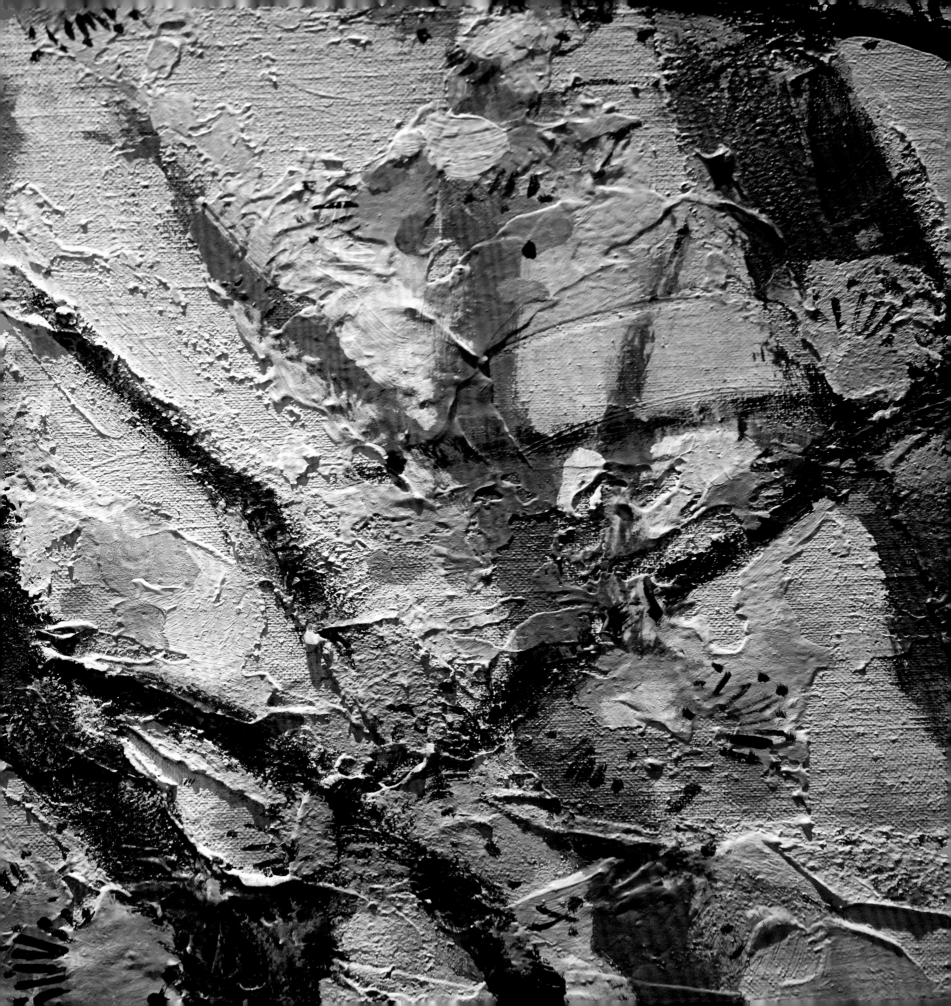

中海·世紀公館銷售会所

The Century Villa 1

中海·世纪公馆别墅 1

Burberry classic card is an easy catch romantic reverie: Sue Glen plaid, unique texture, elegant design, there is After the pattern of time - - - - - - obviously, designers of Burberry A more profound brand interpretation, thus in this project, will be Burberry element clever apply, not dew, classic. The shape Like plus spirit like ness, the spatial pattern of atmospheric fully details with luxury brands Blend, difficult to mask the aristocratic artistically.

BURBERRY 是一张容易引起人浪漫遐思的经典名片：苏格兰格子图案，独特的材质，大方优雅的设计，蕴藏在图案后的时代感-----显然，设计师对BURBERRY有着较为深层次的品牌解读，因此得以在此项目中，将BURBERRY的元素巧妙运用，不露痕迹，经典呈现。形似加上神似，空间格局的大气通透与奢华品牌的细节交融，难掩其贵族气韵。

The Century Villa 2

中海·世紀公館別墅 2

Fresh and tender yellow is work Use very color dot eyeball The color. This is full of the awaken of spring The color of fresh mellow, Combined with white, gray Cool colors, such as the space The colour aamantly Dyed out, thus the whole ivid and lively. Works of Chinese style Birds and flowers, and the phoenix embroidery figure, The Europe type style is mixed Takethe Oriental element, the light Loose and enchanting,added auspicious And the atmosphere andelegant Emotional appeal.

鲜嫩的黄色是作品运用得很点睛的色彩。这种充满春意的新鲜柔美的色彩，再配合白色、灰色等冷色调，将空间的色彩感瞬间渲染出来，从而整体显得生动活泼。作品中点缀的中式花鸟及凤凰绣花图，使欧式的风格又混搭了东方元素，轻松妩媚，增添了祥和的气氛及雅致的情调。

中海·世紀公館別墅 2

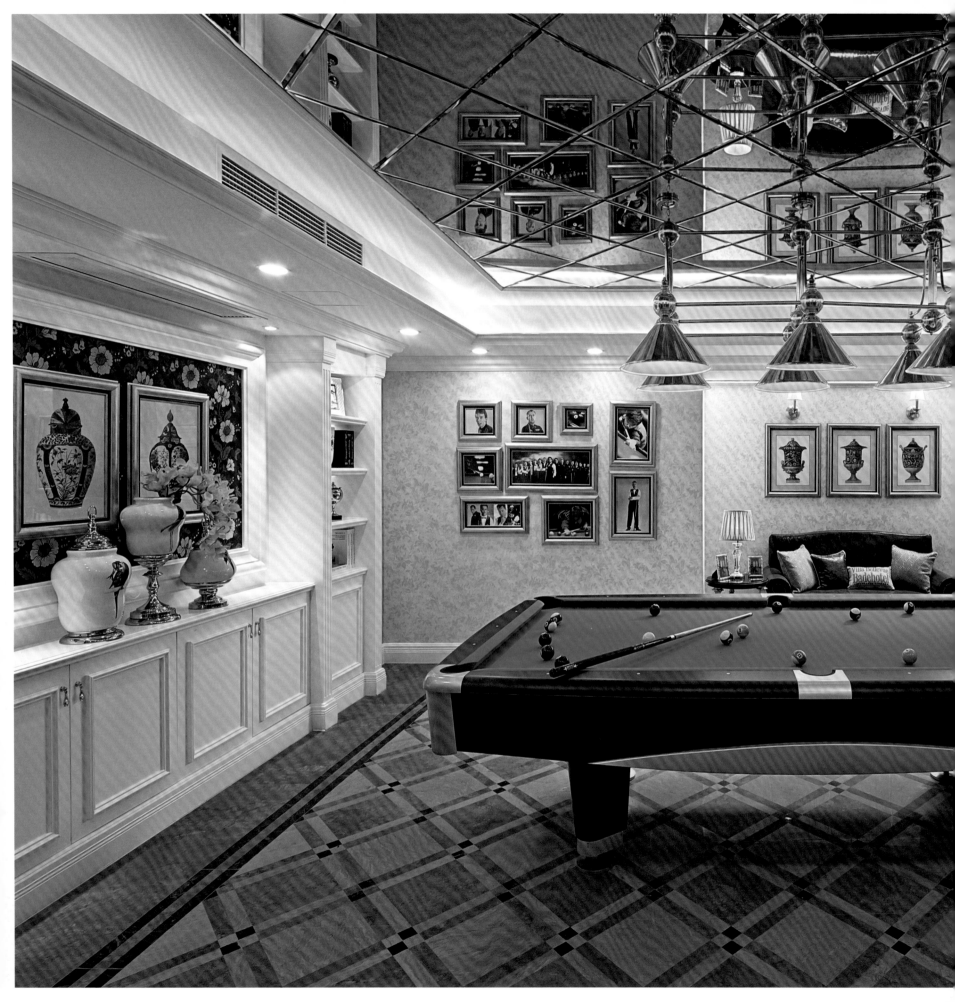

The Aqua Show Flat 1
中海‧碧林灣示範單位 1

Graceful and special: hyacinth. The blue hyacinth is the ancestor of all hyacinth , and Europeans entertain a special feeling for it. Venus in Greek myth was fond of showering with dew on it to promote her skin`s smoothness. In England, used in flowe bunch or decoration flowers, has always playedan indispensable part in the wedding as a symbol of happiness. The blue hyacinth is beautiful but not coquet, delicate but not affected. Compared with light blue, it is deeper; with deep blue, it is purer. Matched with bright light colors, it shows its delicacy and freshness; with dim colors, it exhibits its vitality and wisdom.

风信子,典雅别致。蓝色风信子是所有风信子的始祖。欧洲人对风信子更有一份特殊感情,希腊神话中女神维纳斯,最喜欢以风信子花瓣上的露水来沐浴以使肌肤更为漂亮润滑,在英国,作为幸福象征的蓝色风信子一直是婚礼中新娘捧花或饰花不可或缺的花材。蓝色风信子秀而不媚,娇而不做。这种色彩比之浅蓝,显其深厚;比之深蓝,又显其纯净。搭配高明度的浅色,可以表现出精致清爽;搭配明度低的色彩,则可以表现出精神和知性。

The Aqua Show Flat 2
中海・碧林灣示範單位 2

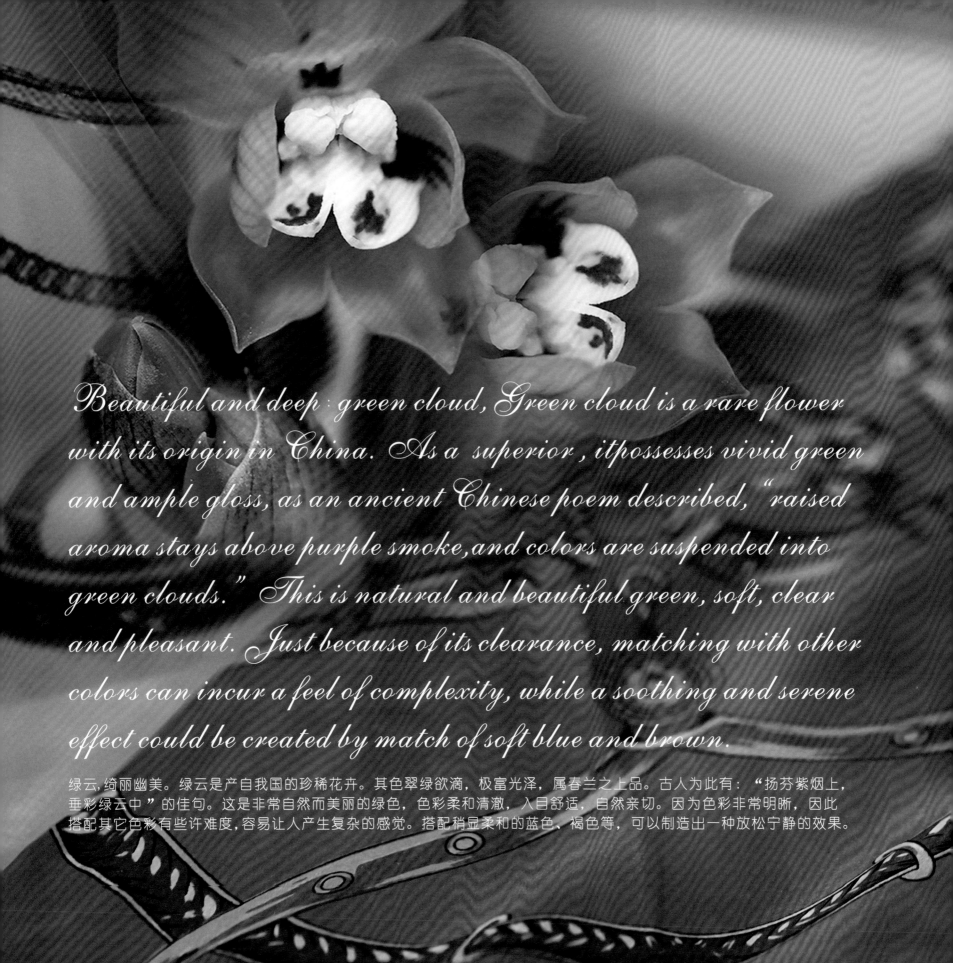

Beautiful and deep : green cloud, Green cloud is a rare flower with its origin in China. As a superior, it possesses vivid green and ample gloss, as an ancient Chinese poem described, "raised aroma stays above purple smoke, and colors are suspended into green clouds." This is natural and beautiful green, soft, clear and pleasant. Just because of its clearance, matching with other colors can incur a feel of complexity, while a soothing and serene effect could be created by match of soft blue and brown.

绿云,绮丽幽美。绿云是产自我国的珍稀花卉。其色翠绿欲滴,极富光泽,属春兰之上品。古人为此有:"扬芬紫烟上,垂彩绿云中"的佳句。这是非常自然而美丽的绿色,色彩柔和清澈,入目舒适,自然亲切。因为色彩非常明晰,因此搭配其它色彩有些许难度,容易让人产生复杂的感觉。搭配稍显柔和的蓝色、褐色等,可以制造出一种放松宁静的效果。

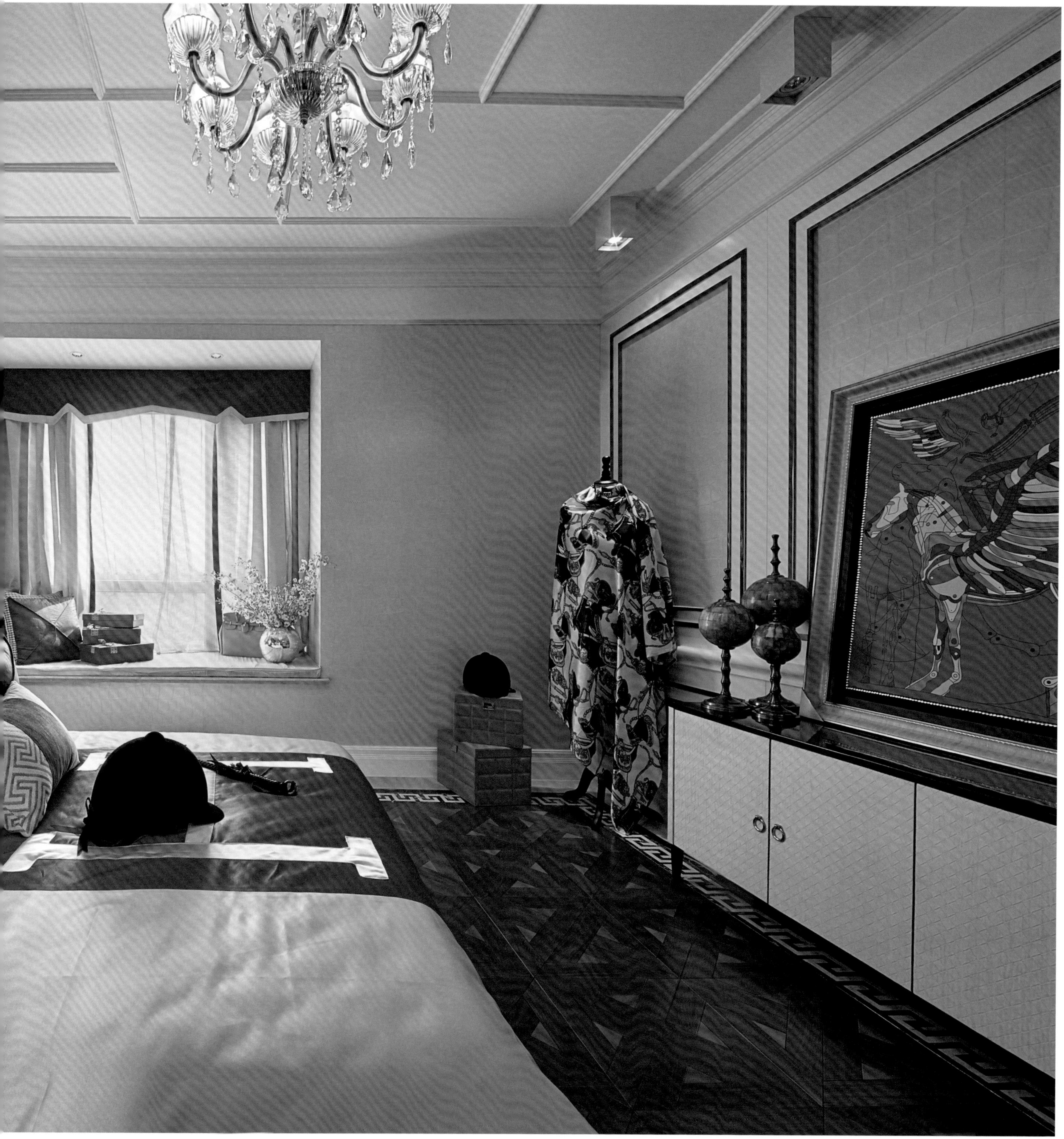

Park Mansion Sales Office

龍湖·世紀景峰銷售会所

Rich and calm: pine cone brown. While brown always implies solemnity, pine cone brown with a slight golden tint is able to make a break through with a light softness and tenderness when displaying nobility, reminding one of the scenery of a distant mountain in the early fall, rich and tranquil, with a profound implication. In the solid color cold gray. background, this color exhibits a strong and ready passion, especially s mart and elegant in response to cold gray.

松果褐，丰足安宁。尽管褐色总是呈现一种庄重而厚实的特点，带着些许浅金的松果褐却能从厚重的感觉中突围，作品在演绎华贵典雅的同时，平添了一份淡淡的柔和与温情，令人不禁怀想初秋远山的景致，丰足而安宁，意境深远悠长。在沉稳的色泽基调下，这种色彩透着一种蓄势待发的热情，尤其在与灰白等冷色系相融合呼应中，它因温暖而尤显灵动优雅。

龍湖 · 世紀景峰銷售会所

龍湖·世紀景峰銷售会所

Crystal Town Show Flat
龍湖・水晶酈城示範單位

龍湖・水晶酈城示範單位

The U World Show Flat
中海・寰宇天下示範單位

Blue Enchantress: Romantic and elegant, Blue Enchantress is a kind of very beautiful and expensive blue rose. With the gorgeous color of sapphire, it can always stand out among numerous colors, showing irresistible charm and also representing deep romance.

蓝色妖姬,浪漫优雅。蓝色妖姬是一种很美且昂贵的蓝色玫瑰。它有着蓝色宝石一般耀目美丽的色彩,在众多色彩中总能脱颖而出,有着无法抵挡的魅力。同时也体现着深深的浪漫情怀。

中海・寰宇天下示範單位

Gate of Peace Show Falt
中海・和平之門示範單位

中海・和平之門示範單位

中海・和平之門示範單位

Glorious City Show Flat
中海・錦城示範單位

Mysterious and noble: tulip. The black tulip is also called Queen of the Night, for its purple in black. One can imagine: when the night falls, the queen travels and unveils herself. What astonishing nobility and beauty it is! Typically, high purity colors like black and dark purple are very suitable for delivering maturity and nobility. Black, which is pure darkness, can accentuate any other color to full extent. That is why we sense strong glamour and beauty from drastic color differences.

郁金香,神秘尊贵。黑色郁金香。又称"夜皇后",黑色中透着微微的紫色。可以遐想:当夜幕降临,"皇后"锦衣夜行,拉开一层神秘面纱,那是怎样的一种令人夺目的尊贵与美丽?通常,在传递成熟、高贵的信息时,运用黑色、深紫色等纯度较强的配色是极为合适的。纯粹的黑色,往往可以最大限度地衬托出任何色彩。因此我们从色彩差异较大的画面中,反而能强烈感受到极具魅惑力的美感。

中海・錦城示範單位

Times City Sales Office
九龍倉・時代上城銷售會所

As a romantic country the national flowers of France, and then it so well, design and color is beautiful, in the group of flowers has a transcendental and free from vulgarity temperament. Legend has it the king of the kingdom of France's first dynasty clovis in to accept baptism, god gave him a present, is the iris. In the design of the case, and then fruitful purple blue petals, large area of green pointed leaves, this as a foil of marigold yellow flower, flower the brick red soil, almost every colour are designers carefully capture.

作为浪漫之国法国的国花，鸢尾花花姿袅娜，花色美丽，在群花之中自有一种超凡脱俗的气质。相传法兰西王国第一个王朝的国王克洛维在接受洗礼时，上帝送给他一件礼物，就是鸢尾花。本案设计中的鸢尾花，丰硕的紫兰色花瓣，大面积绿色的尖头长叶，作为衬托的万寿菊之黄色花蕊，花丛下砖红色的土壤，几乎每一处色彩都被设计师悉心捕捉。

九龍倉・時代上城銷售會所

九龍倉・時代上城銷售會所

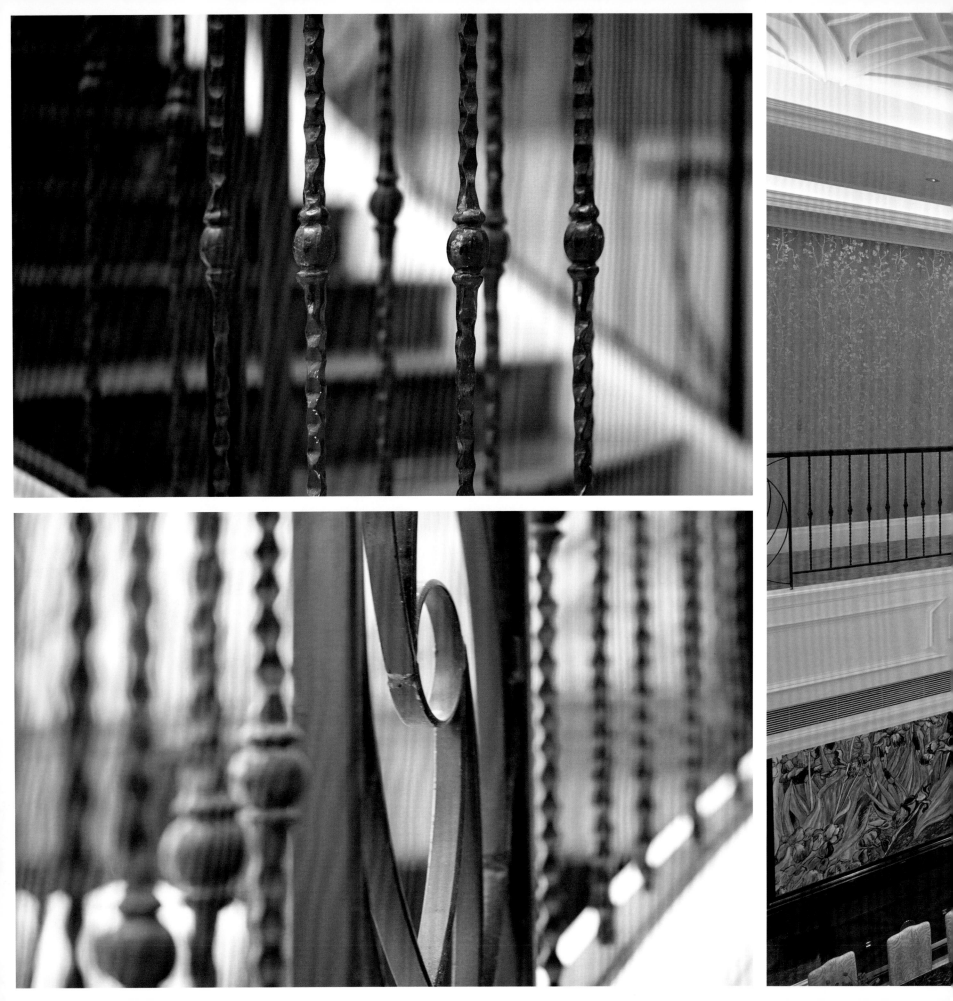

九龍倉・時代上城銷售會所

Times Riverside Show Flat

九龍倉・時代尊邸示範單位

gorgeous and brilliant: Barberton daisy, Originated in South Africa, the Barberton daisy is as a vigorous flower as the sunshine of Africa. Orange chrysanthemums bring red into full play. One cannot but be attracted by its vigor, passion and glory. Still, a cheerful and hilarious scene is created by the match of this bright color and many other colors.

非洲菊,华美亮丽。原产地在南非的非洲菊,如同非洲的阳光一样,是一种能量充沛的花卉。色彩橘红的菊花最大限度地发挥了红色的印象。旺盛的生命力和充沛的激情给人以强大的力量之感,夺目的华贵美丽,令人不禁深深为其陶醉。而这样明丽的色彩,通过与多种色彩的组合搭配,又酿造出欢快、热闹的氛围。

327 九龍倉・時代尊邸示範單位

九龍倉・時代尊邸示範單位

Time Town Show Flat
九龍倉・時代小鎮示範單位

Beautiful and refreshing: white roof iris, The name originates from the Greek language, meaning "the rainbow", which indicates the flower has virtually seven colors. The flower has some sense of nobility as well as freshness. White, with its utmost brightness, is the most capable of manifesting freshness. Light and bright colors are able to create a feeling of extreme thinness, which is the best interpretation of freshness.

白色鸢尾花,优美清新。"鸢尾"之名来源于希腊语,意为彩虹。它表明此类花卉几乎拥有七彩之色。白色鸢尾花洁净清爽之余,略带华贵之感。在所有的色彩中,白色的明度最高,也最能体现清新的效果。浅色调高明度的色彩能营造出薄如蝉翼的感觉,是优美清新印象的最佳诠释。

九龍倉・時代小鎮示範單位

九龍倉・時代小鎮示範單位

九龍倉・時代小鎮示範單位

Sorrento Show Flat
九龍倉・擎天半島示範單位

The refreshing embellishment of vivid yellow accentuates the space originally set in a cold color tone of white and grey with vibrancy and a splash of life. The yellowish hue amiably lightens up the serene dignity and transforms it into shimmers of elegance.

棣棠花。这种鲜嫩的黄色作为白色、灰色等冷色调的点缀,将空间的色彩感一下子渲染出来,生动而活泼。在黄色的点缀下,静谧而有些许庄重的气氛略显轻松,却又增添了几分高雅。

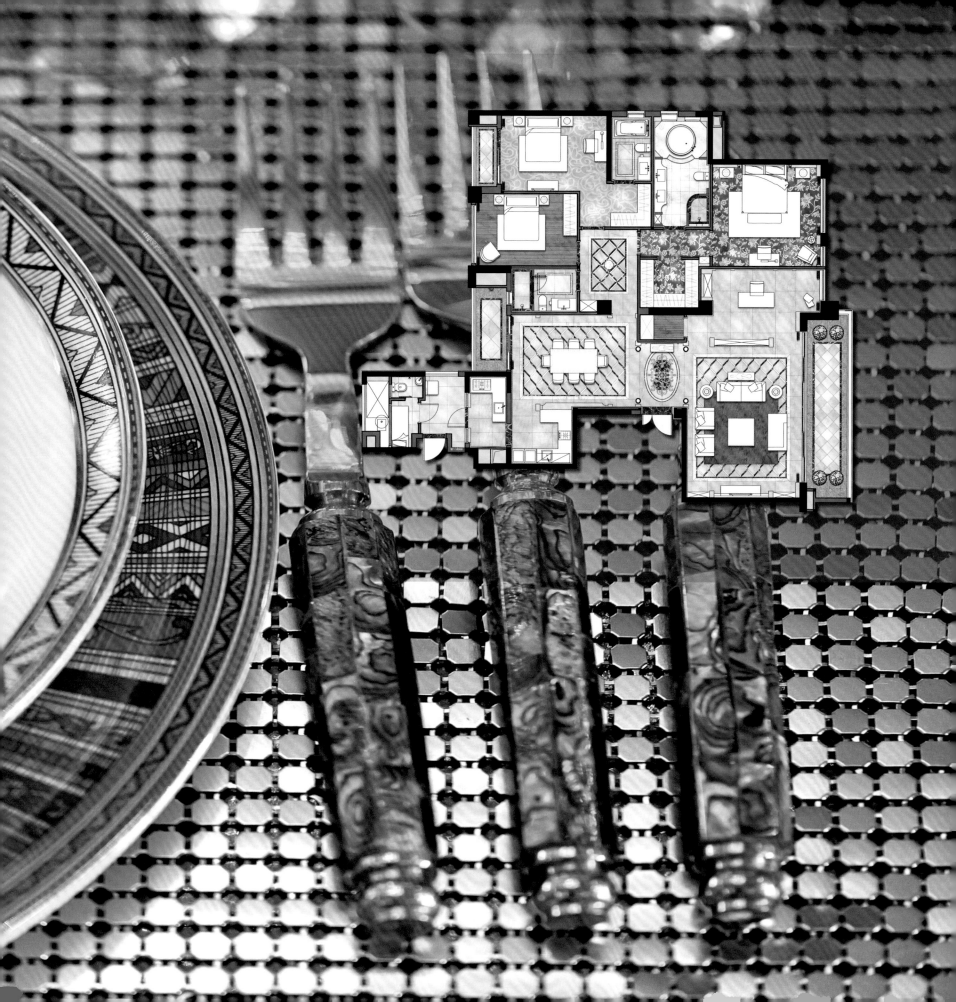

九龍倉 • 擎天半島示範單位

九龍倉・擎天半島示範單位

373 九龍倉・擎天半島示範單位

九龍倉・擎天半島示範單位

377 九龍倉・擎天半島示範單位

HunLiGong Club

華僑城・婚禮宮商務會所

Bright and lofty : Cornflower, The Cornflower has its home in the Europe, especially prevailing in Germany. The blue Cornflower bears a bright and clean color that resemble the sky, creating a feel of serenity and nobility, which would easily captivate bystanders. Compared with the warm colors, a member of the cold color family, blue is typically calm. Like green plants, blue colors have a composing and purifying effect on the mind. An impression of freshness and brightness follows the rising of brilliance and purity.

车矢菊,明净高远。车矢菊的故乡在欧洲,尤其在德国几乎漫山遍野。蓝色的车矢菊有近乎天空一样明净美丽的颜色,酿造出安详而崇高的感觉。这样的蓝色,会让人驻足仰视而不舍离去。与暖色系相比,冷色系中的蓝色往往给人感觉有些冷静和安宁。蓝色系的色彩与绿色植物一样,具有镇静、净化心灵的功能,在提高其明度和纯度后,清爽、明朗的印象也随之而生。

華僑城・婚禮宮商務會所

Success Villa

金雁·領秀別墅

"Dragon can get large or small, rise or hide; it can summon the cloud and frog while getting large, and hide itself and become invisible when getting small; it can also soar into the universe when rising, and lurk itself into great waves when hiding. It has been deep spring now, dragon will change with the opportunity, as if a person achieves his ambition and becomes invincible around the world. Dragon can be compared to the hero of the world." Leadshow private residence villa completed, creatively infused the Chinese element of "dragon" into the classical and European style, so as to reflect the remarkable taste and noble temperament of owners through the extraordinary design.

"龙能大能小，能升能隐：大则兴云吐雾，小则隐介藏形；升则飞腾于宇宙之间，隐则潜伏于波涛之内。方今春深，龙乘时变化，犹人得志而纵横四海，龙之为物，可比世之英雄"。领秀别墅私邸设计，通过巧妙的构思及处理，用中国古典的龙的"意向"将这只"中国龙"简化变身为"龙"的轻巧符号，体现出业主不凡的品位和高贵的气质。

金雁・領秀別墅

LangRun Park Villa
中能·朗潤園別墅

White hibiscus, quiet and calm. Hibiscus toward dusk fell, but every time fade is to open next more gorgeous, its character, with the spirit of even stronger than before and worship Longed for. White hibiscus is elegant and stick to the model. Hibiscus flowers with white as the background color, brown, blue on collocation Quiet composed such as colour, more foil a each respective characteristics and color, build a harmonious and comfortable atmosphere.

白色木槿花,安静沉稳。木槿花朝开暮落,但每一次凋谢都是为了下一次更绚烂地开放,其矢志弥坚的性格,令人敬慕。白色木槿花绝对是优雅而坚持的典范。以木槿花的白色作为背景色,搭配上褐色、蓝色等安静沉稳的色彩,更加衬托出每种色彩各自的特点和美丽,营造出和谐舒适的氛围。

Lang Run International Sales Office
中能·朗潤國際銷售会所

This is Lang Run Park Sales Center, implanting the illusion of "butterfly valley" into reality. The bow tie-shaped door and "butterflies" of various colors gradually flying towards all sides on the outer wall present fresh colors and mystery, attracting people before they enter the lobby. The dominant tone adopts the series of colors of black-white-grey and gold-silver which has no hue, and it looks bright and elegant. Mild light is invisibly set behind each "butterfly", and the second ray of light set "butterflies" off to be more vivid and colorful.

朗润园销售会所,将"蝴蝶谷"的梦幻植入现实。"蝴蝶领结"造型的大门、外墙上渐渐向四方飞去的各色"蝴蝶",色彩清新优美而又具有神秘感。大厅的主色调选取了黑白灰和金银色等没有色相的色彩,明丽高雅,每只"蝴蝶"下方安置了轻盈的提光,二次的光线将"蝴蝶"衬托得更加生动逼真且五彩缤纷。

中能·朗润国际销售会所

443 中能·朗潤國際銷售会所

Acknowledgments 鳴謝

本書得以順利面世,全賴各方的參與與支持,在此由衷感謝
"香港方黃建築師事務所"全體同仁的努力及付出!

于1997年在香港創立的香港方黃建築
建築居住、商業空間整體設計與統籌
過近二十年的發展，從香港到深圳、
國際化設計與管理服務團隊，團隊至
大利、加拿大-----多國多地區的優
以不斷創新務實的設計理念及服務贏
合作客戶既有香港知名房地產開發商
地產……也有國內大型房地產開發企
華潤、中鐵 地產等。項目與作品已遍
天津、成

香港方黃建築師事務所務所，是專業提供多元的營運機構。事務所經相繼成立了高端專業的有中國大陸及香港、意大人與設計師。事務所眾多客戶的支持與信賴。倉、信和、置地、南益海、華僑城、龍湖以及港及廣州、深圳、廈門、重慶等多個城市及地區。

Hong Kong Fong & Wong Architects & Associates

Hong Kong Fong & Wong Architects & Associates (here in after referred to as "Fong & Wong") was established in Hong Kong in 1997, which specializes in integrated design and management of diversified residential and commercial space. After development for over twenty years, besides the head office, Fong & Wong has established high-end professional and international design and management service teams in Shenzhen and Chengdu. Fong & Wong has attracted excellen partners and designers from many countries and areas including China Mainland,Hong Kong SAR, Italy and Canada. By virtue of the design philosophy of constant innovation and pragmatic services, Fong & Wong has won a number of customers' support and trust. The cooperation customers cover both well-known Hong Kong property developers including Wharf, Sino Group, Hong Kong Land, South Asia Real Estate, etc. and famous mainland property developers including China overseas, OCT Properties, Longfor, China Resources Land, China Railway Construction Real Estate Group,etc. Its design projects and works can be seen all over Hong Kong, Guangzhou, Shenzhen, Xiamen, Tianjin, Chengdu, Chongqing and many other cities and areas.

图书在版编目（CIP）数据

设计有约 / 方峻 著 . – 武汉：华中科技大学出版社, 2014.8
ISBN 978-7-5609-9690-5

Ⅰ.①设… Ⅱ.①方… Ⅲ.①室内装饰设计 Ⅳ.①TU238

中国版本图书馆 CIP 数据核字（2014）第185587号

设计有约	方峻 著

出版发行：华中科技大学出版社（中国·武汉）
地　　址：武汉市武昌珞喻路 1037号（邮编：430074）
出 版 人：阮海洪

责任编辑：熊纯	特邀编辑：魏波　王伟	封面设计：张筱雅
责任校对：岑千秀	版式设计：张筱雅	责任监印：张贵君

印　　刷：深圳市彩霸美印刷有限公司
开　　本：889 mm×1194 mm　1/12
印　　张：38
字　　数：228 千字
版　　次：2014 年10月第1版 第1次印刷
定　　价：498.00 元（USD 99.90）

投稿热线：（020）36218949　　duanyy@hustp.com
本书若有印装质量问题，请向出版社营销中心调换
全国免费服务热线：400-6679-118 竭诚为您服务
版权所有　侵权必究